Advocating Shared Responsibilities for Improved Fire Protection

A symposium to examine fire safety challenges of those who cannot take lifesaving action in a timely manner, in the event of a fire, specifically:

Young children (under five),
Older adults (over sixty-five), and
People with disabilities.

CONGRESSIONAL FIRE SERVICES INSTITUTE

900 2nd Street, N.E.
Suite 303
Washington, D.C. 20002
(202) 371-1277
FAX (202) 682-FIRE
Website: www.cfsi.org

Senator William Roth
Honorary Chairman

Senator Richard Bryan
Honorary Co-Chairman

Senator John McCain
Honorary Co-Chairman

Senator Paul Sarbanes
Honorary Co-Chairman

Congressman Robert Andrews
Honorary Co-Chairman

Congressman Sherwood Boehlert
Honorary Co-Chairman

Congressman Steny Hoyer
Honorary Co-Chairman

Congressman Curt Weldon
Honorary Co-Chairman

William Jenaway
President

Louis J. Amabili
M.H. Jim Estepp
Arthur Glatfelter
Chet Henry
Board Members

William M. Webb
Executive Director

September 20, 1999

Greetings:

To the individuals and organizations associated with Solutions 2000, I speak on behalf of the Congressional Fire Services Caucus when I say thank you for bringing national awareness to an important fire safety challenge.

The following report represents a new direction towards protecting at-risk populations from the threat of fire – a threat that accounts for over 4,000 deaths annually and over 23,000 injuries. Whereas many individuals are capable of protecting their personal well-being, physical limitations attributed to age and disabilities impair the mobility of others, placing them at a higher risk.

I commend Solutions 2000 for their dedication and commitment to this challenge. The report embraces a comprehensive and integrated approach that addresses the disciplines of education, engineering and enforcement. They are not addressed separately but rather interdependently with each discipline playing an equally important role. Changing public attitude about education, enforcement and engineering can be a time consuming process. But the synergy of these organizations will dictate the pace of progress and I have every confidence that they will succeed.

Many members of Congress are already addressing some of the issues contained in this report. I am a strong advocate of fire resistant sleepwear for our children, and continue to support legislative efforts to require the Consumer Product Safety Commission to restore the original flammability standard for children's sleepwear. Other members have championed the causes of fire safe cigarettes, flame-resistant upholstered furniture, and the use of built-in protection such as automatic fire sprinklers to protect lives and property from fire.

This report is a blueprint for protecting at-risk populations against fire. In it, there is a role for everyone who has a vested interest in fire and life safety. I hope others will join us in this important crusade to address the challenges facing individuals in need of our support.

Best wishes for your continued success and safety.

Sincerely,

William V. Roth, Jr.
Chairman
Congressional Fire Services Caucus

Solutions 2000

A Symposium To Examine Fire Safety Challenges Of Those Who Cannot Take Life Saving Action In A timely Manner, In The Event Of Fire.

North American Coalition For Fire and Life Safety Education

American Red Cross

British Columbia, Canada, Fire and Life Safety Advisory Committee

International Association of Fire Marshals

National Association of State Fire Marshals

National Fire Information Council

National Fire Protection Association (NFPA)

National Fire Sprinkler Association

Operation Life Safety

United States Fire Administration

Symposium Planning Team

AARP

American Association of People With Disabilities

American Red Cross

Congressional Fire Services Institute (CFSI)

Eastern Paralyzed Veterans Association

International Code Council, Inc.

National Fire Protection Association (NFPA)

National Fire Sprinkler Association

National SAFE KIDS Campaign

Paralyzed Veterans of America

September 20, 1999

To Readers of the Solutions 2000 Report:

In April of 1999, the North American Coalition for Fire and Life Safety Education conducted a symposium to examine fire safety challenges of those who cannot take life saving action, in a timely manner, in the event of a fire; specifically young children (under five), older adults (over sixty-five), and people with disabilities. The symposium was conducted in Washington, D.C. in conjunction with the annual Congressional Fire Services Institute Dinner.

The symposium brought together preeminent experts representing fire safety and the specific concerns of young children, older adults, and people with disabilities. The participants spent two intense days exploring present and futuristic practices and interventions for providing improved fire safety for those less able to protect themselves from unwanted fire.

The following report represents the suggested "solutions" formulated by the attendees. It is important to note that they intend to address "shared responsibilities" of both fire service representatives and representatives of the advocacy groups. Therefore, it is our fervent desire that fire safety experts and experts representing these high-risk groups, continue to work together in the future to implement the recommendations.

I congratulate the members of the North American Coalition for their willingness to conduct the symposium and wish to thank our speakers for helping to set the stage. In addition, I would like to thank Jim Dalton and the Planning Team who gave much time and effort to ensuring the success for the Solutions 2000 symposium. Lastly, we would be remiss if we did not express our sincere thanks to the National Fire Protection Association (NFPA), its Center for High Risk Outreach, the National Fire Sprinkler Association (NFSA), KIDDE International, the American Association of Retired People (AARP), and FEMA/USFA for their generous financial and in-kind support.

Sincerely,

Rocky Lopes

Rocky Lopes, Ph.D.
Chairman
North American Coalition for Fire and Life Safety Education

TABLE OF CONTENTS

SOLUTIONS 2000 PLANNING TEAM

Jim Dalton
Chair of Solutions 2000 Planning Team
Director of Public Fire Protection
National Fire Sprinkler Association, Inc.
(NFSA)

Meri-K Appy
Vice President of Public Education
National Fire Protection Association
(NFPA)

Kim Beasley
Director of Architecture
Paralyzed Veterans of America
(PVA)

Helena R. Berger
Executive Director
American Association of People with Disabilities
(AAPD)

Brian D. Black
Director, Building Codes and Standards
Eastern Paralyzed Veterans Association
(EPVA)

Rocky Lopes, Ph.D.
Chair, North American Coalition for Fire and Life Safety
Education
Community Disaster Education
American Red Cross

Hank Fellner
Fire and Burn Team Leader
The National SAFE KIDS Campaign

Leon Harper
Manager of Housing Programs
American Association of People with Disabilities
(AARP)

Soy L. Williams, A.I.A.
Government Relations Director
International Code Council, Inc.

Sara C. Yerkes
Director, Government Affairs
National Fire Protection Association, Washington Office
Liaison to Congressional Fire Services Institute

Management of Solutions 2000 Provided by:
Peg Carson
Carson Associates, Inc.
Fire Protection Consultants
Warrenton, Virginia 20186

This publication was produced under contract by TriData Corporation for the United States Fire Administration, Federal Emergency Management Agency. Any information, findings, conclusions, or recommendations expressed in this publication do not necessarily reflect the views of the Federal Emergency Management Agency or United States Fire Administration.

EXECUTIVE SUMMARY

In April of 1999, a group of North America's foremost authorities on fire protection and prevention convened with experts who represent three sections of the population with abnormally high fire risks: young children under the age of five, older adults over the age of sixty-five, and people with disabilities. *Solutions 2000* marked an unprecedented initiative to explore the fire safety needs and challenges of those who cannot take lifesaving action in a timely manner in the event of a fire. *Solutions 2000* represented the collaborative efforts of various organizations, associations, industries, educators, and individuals, not exclusive to the fire service, to address the multitude of fire risks facing these groups.

Solutions 2000 Objectives

- Bring together preeminent experts representing fire safety and the specific concerns of young children, older adults, and people with disabilities.

- Provide fire safety statistical information pertinent to the target groups.

- Explore state-of-the-art practices and interventions for providing fire safety for those less able to protect themselves in the event of fire.

- Explore new practices and interventions and recommend a plan of action addressing the shared responsibilities for improving fire safety among the target groups.

- Publish and disseminate symposium findings and recommendations including priorities and steps for implementation.

The goal of *Solutions 2000* was simple – to propose ideas and initiatives to reduce or eliminate fire-related casualties in young children, older adults, and people with disabilities. The logistics involved in reaching this goal, however, are complex. Each conference participant was charged with identifying possible barriers and developing realistic means to overcome them. To develop effective strategies that target the fire problem, participants used an integrated approach aimed at creating realistic solutions. These solutions encompassed the disciplines of education, engineering, and enforcement. Developing solutions in all three disciplines is collectively known as the "systems approach;" it is a philosophy that views diverse problems from a variety of perspectives. With this approach, the attendees of *Solutions 2000* tried to tackle a huge task in a relatively short amount of time. As a result of their dedication, wisdom, and foresight, we are now able to produce a document that will not only benefit the fire service, but those who represent and serve the needs of these special populations.

Key Solutions

The following briefly summarizes key ideas that arose in the symposium for each target population. The body of the report presents a more detailed explanation of the ideas raised. Many of the ideas raised for one group apply to the others and, in some cases, to the general population.

Young Children

- Form a coalition that focuses on child fire safety and awareness.

- Develop fire safety programs specifically focused towards children with disabilities.

- Prepare children for fire emergencies by getting parents, siblings, caregivers, educators, and role models involved in a child's fire safety education.

Older Adults

- Promote life safety, not just fire safety, in programs addressing older adults.

1

- Do not isolate or single out older adults in fire safety programs; older adults prefer mainstream messages that apply to all sections of the population.[1]

- Expedite the development of "smart stoves[2]" (cooking is the leading cause of elderly fire injuries).

People with Disabilities

- Educate the fire service and building design community on fire safety considerations for people with disabilities.

- Organize disability and fire service representatives into a national coalition with two goals: to raise fire safety awareness among the disability community; and to raise the awareness of the fire service to the needs of people with disabilities.

- Improve emergency egress from buildings that house people with disabilities.

- Form a coalition to expedite the implementation of the fire safe elevator.[3]

- Pay more attention to fire safety issues of people with disabilities during the code development and enforcement process.

---

[1] There is a minority opinion that this is not a good idea, and that the elderly groups should be singled out and targeted with special materials.

[2] A "smart stove," also referred to as a fire safe stove, is designed to shut itself off before the food starts to burn, and thus cause the potential for fire. This may not cure the problem of dangling the sleeves of loose fitting garments over stoves. Fire safety education must stress the importance of the proper clothing during cooking.

[3] A fire safe elevator is an elevator that can be used for safe egress during a fire emergency without firefighter control. The technology is available, it simply needs to be applied.

Universal Messages

- Promote the installation of home fire sprinklers, make the costs more affordable, and educate the public on the benefits of fire sprinkler systems in general.

- Form a coalition that will make fire safety a primary concern by raising our safety expectations for the environments to which young children, older adults, and people with disabilities are exposed.

- Mandate more built-in fire safety in new construction; it is less expensive to install fixtures to outfit a structure during its initial construction phase, than it is to later retrofit.

- Promote life safety, not just fire safety, in programs addressing all audiences.

I. INTRODUCTION

On a per capita basis, young children and older adults have a greater propensity for being injured or killed in a fire. Children accounted for 12% of all fire related deaths in the United States during the period 1987-1996, while older adults over the age of sixty-five accounted for 26% of all deaths during the same period. Although not quite as dramatic, the injuries suffered by these populations are also high for the period 1987-1996; children accounted for 5% of all injuries while older adults accounted for 12% of all fire-related injuries. There is, however, little data available on the associated fire risks for people with disabilities. People with disabilities are presented with a greater challenge for detection of, and escape from, a fire. Young children and older adults also are unable to react to a fire emergency in the same fashion as the remainder of the population. Young children, by virtue of their age, innocence and lack of judgment, are often the victims of fire simply because they cannot or do not know enough to remove themselves from the danger. Diminished physical abilities and senses associated with aging expose older adults to a multitude of fire risks. This population is limited in its ability to detect and escape a fire and, hence, is more likely to sustain injury. People with disabilities have significant fire risks as well. Sensory impairments may inhibit the disabled person's ability to detect a fire, while physical impairments may inhibit life-saving evacuation.

Three years ago, the North American Coalition for Fire and Life Safety Education held a symposium in which these three high-risk groups were identified as needing more attention in fire safety. The symposium found that the fire service does not often act as advocates for young children, older adults, or people with disabilities when fire safety is called into question – or at least do not give them adequate support. In recognizing fire and life safety personnel as instruments of change, it is possible to reach the hard-to-reach, to teach the hard-to-teach, and to promote better engineering and wider enforcement. It is from this impetus that *Solutions 2000* emerged as a means to strengthen problem-solving techniques and especially to incorporate the opinions of those with a vested interest in serving the needs of high-risk populations.

Representatives from 54 agencies and organizations met in Washington, D.C. on April 20 and 21, 1999 to discuss the inherent dangers of fire and the challenges in fire safety for the three target populations. Fire service representatives were invited to the conference based on their knowledge in the areas of education, engineering, and enforcement; these representatives included fire chiefs, local and state fire marshals, public educators, fire protection engineers, and fire data analysts. To guide and enhance the potential findings of the conference, members from a variety of advocacy groups who represent young children, older adults, and people with disabilities were invited. These experts' insight into persistent fire safety threats for each of the three populations was vital in not only identifying current barriers to fire

safety, but also in identifying problematic side effects to traditional solutions. Consumer advocacy groups and a variety of non-profit public education and life safety organizations were another subset of conference attendees. Representatives from state, local, and federal government agencies involved in public health were also asked to participate. Dr. Rocky Lopes, American Red Cross, who serves as the Chairman for the North American Coalition for Fire and Life Safety Education, served as the principle facilitator for all activities of *Solutions 2000*.

The recommendations contained herein are the outcome of a collective process that included individuals who represent 54 organizations. The recommendations do not necessarily represent the sole opinion of any specific group.

II. METHODOLOGY

The Three E's: Education, Engineering, and Enforcement

To effectively address the fire safety needs of any population, the three E's, education, engineering, and enforcement, must be addressed. There are certain fire risks that may be best addressed through educational efforts, while others may be better served by increased enforcement or engineering techniques. Each of the three E's exerts a synergistic effect on the others, however, and together they are much more effective than individually. Education can be used to promote engineering possibilities, such as home fire sprinkler systems. Code enforcement can be used as an opportunity for education. Point-of-sales information tags can tell consumers how to use the safety features engineered into products. Each of the three E's can contribute to the development of comprehensive, realistic, and effective solutions. Collectively, they can reduce the effects of fire, if not prevent them.

The three E's guided *Solutions 2000* participants during their proceedings. The findings of the conference, however, are not restricted or divided into the categories of education, enforcement, and engineering.[4]

Conference Group Approach

Conference attendees were assigned to one of three conference groups designated to address one of the following population groups: young children, older adults, and people with disabilities. Over the course of two days, each conference group held four separate discussion sessions during which they discussed fire concerns and solutions relevant to their target population. Each group was given the challenge of identifying concerns and developing solutions to the fire problem as it pertained to their target population.

During Session 1, group members examined current conditions, barriers, and challenges in fire safety. Using a series of opening presentations on relevant statistics and the education, engineering, and enforcement approaches to fire safety as building blocks, participants discussed topics they found particularly intriguing or disturbing, or of special relevance to their job. Session 2 was dedicated to the generation of ideas and innovative thinking about fire safety. After identifying current barriers and problems, as well as possible solutions, Session 3 was targeted towards creating realistic action plans in the areas of education, engineering, and enforcement. The action plans outlined the leaders, partners, and short and long-term goals of the solution strategy. Once developed, participants used Session 4 to discuss their solutions to determine which were the most plausible, and the best way to implement them.

---

4 Editor's Note: We chose not to divide the responses of the break-out groups into the education, enforcement, and engineering categories as there was so much overlap and synergy between the ideas and solutions.

The conference concluded with a summary report from each of the three assigned population groups that outlined the results of all four discussion sessions. Each group was given the opportunity to merge their accomplishments in the fields of education, engineering, and enforcement into a comprehensive, action-oriented plan and present recommendations for implementation. The comprehensive, action-oriented statements are presented in Section IV of this document.

III. BACKGROUND DATA ON THE FIRE PROBLEM IN THE UNITED STATES

To start the conference, Dr. John Hall provided a statistical overview of the fire problem for each of the three target groups. Other speakers introduced the state-of-the-art in education, engineering, and enforcement as it pertains to the target groups. Dr. Hall's presentation, which included extensive fire data, as well as examples of solutions to the fire problem using the three E's, is reviewed below. The synopses for the remaining three presentations can be found as Appendices A-C.

"Framing the Problem"
– John R. Hall, Jr., Assistant Vice President, Fire Analysis and Research,

National Fire Protection Association, Quincy, MA

As the Assistant Vice President of the Fire Analysis and Research Division of the National Fire Protection Association, Dr. Hall is one of the premiere national experts on fire data. The following summarizes Dr. Hall's outline of the problem that fire poses in the United States.

In 1997 alone, the United States experienced 1,795,000 fires, killing 4,050 civilians and injuring another 23,750. The direct property damage alone caused by these fires totaled $8.5 billion. The total cost of fire, including indirect costs such as fire personnel, medical expenditures, insurance overhead, built-in protection, and attributed cost of deaths and injuries is estimated as high as $205 billion.[5]

Prevention

Preventing a fire is the ultimate goal of fire safety officials. There are three means by which this may be accomplished:

- Change the heat source

- Change the fuel source

- Change the behavior

By limiting any one of these factors, it is possible to prevent a serious, even fatal, fire. Using a systems approach, Dr. Hall analyzed some of the leading causes or scenarios of fire, and the engineering, education, and enforcement approaches to limiting or removing the causative factors.

Smoking Materials. Fires caused by smoking materials are the leading cause of fire deaths in the nation, accounting for 27 percent of the total deaths. This cause should not be called "careless smoking," because such a label inappropriately prejudges that the behavior is the only factor in the fire cause. In fact, there are alternatives in engineering and enforcement to address fires caused by smoking materials.

Engineering:	Design an ignition-resistant cigarette
Education:	Teach smokers to be more careful
Enforcement:	Strictly enforce no-smoking regulations.

Upholstered Furniture. Approximately 18 percent of fire deaths begin with ignition of upholstered furniture. Ignition can occur in a number of ways, including from an open flame source, such as a cigarette, or radiant heat from nearby equipment. From a systems viewpoint, there are a number of means by which this type of incident may be avoided.

Engineering:	Make upholstered furniture resistant to open flames
Education:	Teach the public to keep furniture away from open flames and potential heat sources
Enforcement:	Make the Upholstered Furniture Action Council recommendations regulations

Arson. Arson is the second leading behavior or heat source cause of fire deaths, accounting for one-sixth of the total. Particularly noteworthy is the age profile of arsonists; approximately 1/2 of all people arrested for arson are under the age of 18.

Engineering:	Install better security devices
Education:	Counsel juvenile firesetters
Enforcement:	Investigate all fires of unknown cause to better identify arson.

Mitigation[6]

Despite our best prevention efforts, fires still occur. Once this occurs, what one can do is try to mitigate its impact. There are several approaches to mitigation, including the following:

- *Limit Fuel Loads.* Most deaths occur in post-flashover fires. To prevent this, make it

---

[5] The latest fire data may be obtained by contacting the NFPA One-Stop Data Shop.

[6] The definition of mitigation used by emergency managers does not include rapid detection and notification.

harder to, or make it take longer for fire to, reach the flashover point.

- *Rapid Detection and Notification.* Twenty-five years ago, very few homes were equipped with smoke alarms. Today, less than 7 percent of homes do not have one. However, 42 percent of reported fires and 59 percent of fire deaths occur in these homes. In addition to homes without smoke alarms, approximately 1 in every 5 homes with alarms have alarms that are not functioning. One-third of all home fires in homes with alarms are in homes with non-functioning smoke alarms.

- *Rapid Suppression.* Less than 1 percent of homes experiencing a fire are equipped with a fire sprinkler system. More apartments, especially high-rises, experiencing fires have fire sprinkler systems, but their numbers can be improved. Fire sprinkler systems are very effective and may cut fire deaths by one-half to two-thirds in properties where they are installed.

- *Compartmentation.* Internal barriers help halt the spread of a fire and confine it to the room of origin. The construction type is correlated with the extent and type of internal barriers, and so is correlated with the probability of flame spread.

- *Evacuation.* Although everyone should practice a fire escape plan, most households do not.

Table 1 shows how each of these fire approaches can be implemented using the three E's. Mitigation may be enhanced in much the same way as prevention by altering one or all of the three E's.

Table 1. Solutions to Improve Mitigation

	Education	Engineering	Enforcement
Limit fuel loads	Reduce clutter	Design room surfaces to have lower flame spread ratings	Disallow indoor storage of gasoline
Rapid detection and notification	Test and maintain smoke alarms	Hard-wired smoke alarms w/ battery backup, long-life batteries, hush buttons	Require detection/alarm systems
Rapid suppression	Teach the value of a fire sprinkler, dispel myths	Make fire sprinkler systems more affordable	Require fire sprinklers in all new homes and create retrofitting incentives
Compartmentation	Close doors, keep fire doors and exit doors closed at all times	Install automatic door closers	Require door closers; high-rise inspection by FD, building managers
Evacuation	Plan and practice an escape plan; know the layout of the home	Install panic hardware to alert members of household	Mandate fire drills (public places, apartment and group homes)

Risk Factors for Fire Deaths

Certain demographic and behavioral characteristics expose members of the population to unduly high fire death risks. The leading risk factors for fire deaths are as follows:

- *Age (Home Fire Death Rates vs. Age (Chart))*. Fire death rates per million population are greatest for people under age 6 and over age 65. Age is an un-modifiable risk factor, but steps can be taken to protect young children and the very old from the vulnerability associated with their ages.

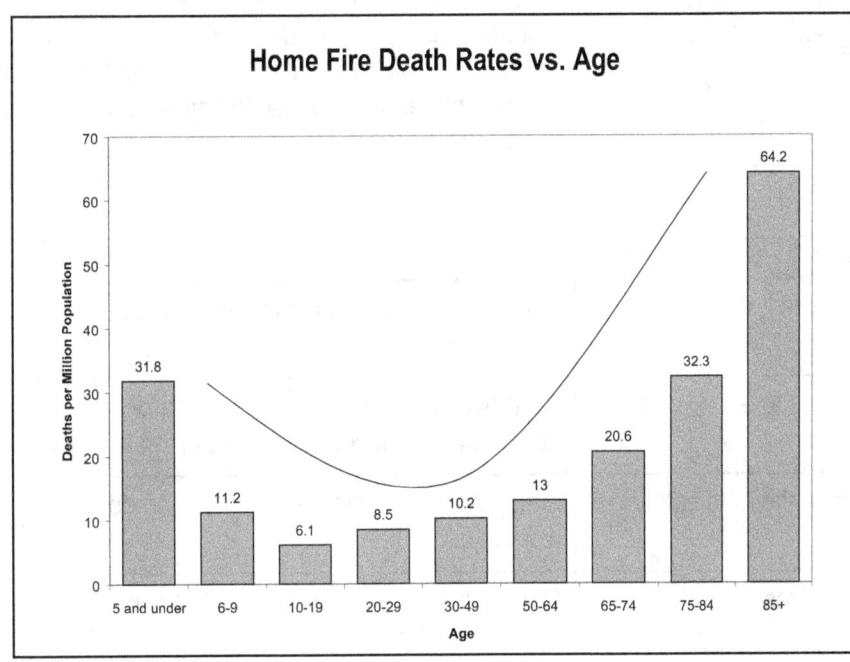

- *Disabilities and Impairments*. The current reporting system undercounts victims of fires with a disability or impairment. The following data from NFIRS, therefore, should be considered minimums:

  - 10 percent are under the influence of alcohol or drugs

  - 8 percent have the physical or developmental limits of young children

  - 7 percent have a physical disability

  - 3 percent have the physical or mental limits of older adults

  - 2 percent have a mental disability

Some fire victims have multiple impairments, but this data can only capture single impairments. The use of alcohol or drugs is the only condition on this list that is preventable. Alcohol and drugs are a factor in about 40-50 percent of adult home fire deaths. They are factors in approximately 25-33 percent of home fire deaths of all ages.

- *Poverty and Related Factors*. Poverty and education levels may explain up to 1/3 of the variation in state fire death rates. For example, in the State of Louisiana, 1/4 of the population is below the poverty line, and they have a fire death rate of 25 per million population. In contrast is the state of New Hampshire, where only 1/13 (7 percent) of the population resides below the poverty line; New Hampshire's fire death rate is 7 per million population. In Kentucky, 25 percent of the adult population did not finish high school; the fire death rate there is 23 per million population. However, in Wyoming, where only 9 percent of the adult population did not finish high school, the state averages 5 fire deaths per million population.

- *Where You Live.*

International Comparisons – The United States and Canada have two of the highest fire death rates in the Western industrialized world. Current theories attribute much of the difference to cultural attitudes and associated behaviors and infrastructures surrounding fire in different nations. Although other nations have lower fire deaths, some trends appear to be changing. For example, in the United Kingdom, the incidence of arson is increasing, while in the United States it has been decreasing. The United States also has the greatest number of home smoke alarms per million population in operation. Overall, the fire death rate in the U.S. has decreased more over the past two decades than in other Western nations, and is nearly equal to the rates in some European and Pacific Rim countries. But for now, we are still among the highest.

Fire Death Rates Also Vary By Community Size – Communities where the population is less than 2,500 or greater than 500,000, have the highest fire deaths per million population. This may be related in part to the poverty associated with both rural communities and urban cities. Poverty may also explain why the Southeastern region of the United States, which is also the poorest, has a higher fire death rate than other regions in the country.

Perceptions vs. Reality

Sometimes people place themselves at unnecessary risk for death from a fire. Common misconceptions about the dangers of fire may prevent people from taking immediate, life-saving action.

Perception	Reality
People think they are the safest from fire in their homes	The risk of dying in a fire is greatest in one's home and in one's car.
You have at least 10 minutes to escape from a fire.	Flashover can consume a room and kill occupants throughout the home in 2-4 minutes.
Males are more confident in their fire safety.	Men have a 50% greater fire risk than do women do.
Older adults feel safe from fires.	Older adults are one of the most at-risk populations for dying or being injured in a fire.

Suspected Risk Factors that Aren't

There are several mythical fire death risk factors. Although they pose no threat alone, their statistical association with known risk factors causes them to be viewed as risk factors on their own. Upon closer examination, we see that they pose no significant risk independently.

Associated Myth	Truth
Race	Race may appear to be a factor, but on closer examination, race is closely linked to poverty, a documented fire risk factor that is better correlated with risk than race is.[7]
Age of housing	The age of a house may appear to be a risk factor, until one adjusts for the poverty of occupants of much older housing. However, older electrical systems are a risk factor.
High-rise buildings	High-rise buildings are less of a fire risk than low-rise buildings as they have more built-in protection.
Portable kerosene heaters	Gas-fueled space heaters are riskier than portable kerosene heaters.
Manufactured homes	Manufactured homes built after 1976, following the implementation of HUD standards, have the same fire risk as site-built dwellings.
Latchkey children are at high risk	The lack of supervision that leads to fires involving children is usually of another type, i.e., not due to the absence of parents at work.

Conclusion by Dr. Hall

We can successfully reduce fire casualties through either prevention or mitigation. Improving education and engineering are options for developing solutions to every problem. In more severe cases, efforts to improve enforcement are necessary to implement change. Enforcement is about covering everybody and ensuring that the right things are being done. It is important to remember, however, that you don't always need a law to enforce fire safety. Anyone with a vested interest in the health and well-being of these special groups can act as an agent of education, engineering, and enforcement.

IV. SOLUTIONS PROPOSED BY EXPERT GROUPS

A number of proposed solutions were offered by the three expert groups at the conclusion of the conference. Each of the groups focused on the three E's: education, engineering, and enforcement, and how to relate them to their subject populations. The key ideas for each group, in addition to a section entitled "Universal Messages," for those ideas that were stressed in multiple groups, are discussed below.

Children

Form a coalition that focuses on child fire safety and awareness.

---

[7] There are, however, some ethnic correlations with risk even after adjusting for poverty.

- Develop the coalition from a broad spectrum of groups that have real or potential influence on fire safety for children. The coalition should include representation from the fire service, children's organizations, disability groups, and service organizations. Some of these groups include United Cerebral Palsy, Centers for Disease Control and Prevention, American Red Cross, Self Help for Hard of Hearing, National Association for Disabilities, Lions, Paralyzed Veterans of America, National Fire Protection Association, National SAFE KIDS Campaign, Juvenile Firesetting Intervention Programs, American Academy of Pediatrics, and Emergency Medical Services for Children.

- Influence and motivate groups and agencies (e.g. Consumer Product Safety Commission) that can advance technical fire safety solutions for children (e.g. safe lighters and fire resistant sleepwear).

- Select a task force from the coalition to conduct a thorough review of the current materials and programs available to identify where gaps in fire prevention and education for children exist.

- Develop new materials and programs while simultaneously restructuring existing programs to bridge the gaps identified.

- Identify or develop ways to ensure that children, parents, caregivers, and the fire service are familiar with child-specific fire prevention materials, including pamphlets, videos, role-playing, and juvenile fire-setter prevention programs.

Develop fire safety programs focused towards children with disabilities.

- Identify the fire safety problems specific to various cultural and developmental differences and physical and mental disabilities. The disabilities addressed during fire safety education should at least include ADD/ADHD (Attention Deficit Disorder / Attention Deficit Hyperactivity Disorder), mobility impairments, hearing and vision impairments, and cognitive processing disorders.

- Develop programs and materials appropriate to each of the above groups.

- Identify and motivate the appropriate agency or coalition concerned about particular types of disabilities to help develop these materials. Encourage these groups to help deliver the programs to children in their constituency.

- Install an evacuation chair in homes with children with physical disabilities, where appropriate.[8]

Prepare children for fire emergencies by getting parents, siblings, caregivers, educators, and role models involved in a child's fire safety education.

At a minimum:

- Support the delivery of comprehensive fire and life safety education in schools.

- Provide organizations that serve and interact with children and their caregivers with fire and life safety information and materials.

- Teach children how to react appropriately in the event of fire.

- Teach children how to call the fire department in case of an emergency. Not every jurisdiction uses the *911* system.

---

[8] An evacuation chair is a mechanical device that runs on a track up and down a stairwell. It can be equipped with either a platform for a wheel chair or a built-in chair.

- Teach children a fire escape plan from every room in the house; they should know at least two ways out of each room. These drills should be practiced often, including in the dark.

- Teach children not to be afraid of firefighters.

- Include the fire department in fire drills and pre-fire planning.

- Alert firefighters to the presence of small children and children with disabilities and special needs before a fire. An emergency communications center (*911,* for example) database reporting system could be developed to inform firefighters at the time of dispatch of a potential special rescue.

- Promote greater use of fire sprinkler systems, especially in schools, day care centers, and homes. Fire sprinklers can save young children who cannot escape on their own.

Older Adults

Promote life safety, not just fire safety, in programs addressing older adults.

- Identify the areas in fire safety that are lacking specifics for older adults.

- Combine the expertise of the fire service industry and older adults advocacy groups (e.g. AARP) to develop fire prevention and education programs, in addition to an effective way to market the new programs.

- Include life safety education in the materials and programs for fire safety.

- Encourage the fire service to collaborate with advocacy groups to expand the outreach of its message and promote the well-being of older adults.

Do not isolate or single out older adults in fire safety programs; older adults prefer mainstream messages that apply to all sections of the population.[9]

- Follow the advice of recent market research studies that shows older adults do not want to be singled out; many feel that it supports a stereotype that the elderly are all frail and helpless, which is not their self-image and using that image may turn many off from fire safety messages.

- Develop universal fire safety messages that pertain to all sections of the population.

- Solicit advocacy groups to add their own subtleties to universal fire safety messages for their older constituents. Test, market, and package these materials.

- Promote fire sprinkler systems for all homes and for all care institutions; fire sprinklers can help save older adults, who are the age group at highest risk.

Expedite the development of "smart stoves."[10] Cooking has been identified as the leading cause of fire injuries to older adults.

- Seek funding to continue research and prototype development of the "smart stove."

- Solicit support from agencies, such as NIST, Underwriter's Laboratories, and CPSC to test and prove the safety and efficacy of the "smart stove."

---

[9] There is a minority opinion that does believe older adults should indeed be targeted with special materials.
[10] A "smart stove," also referred to as a fire safe stove, is designed to shut itself off before the food starts to burn, and thus cause the potential for fire. This may not cure the problem of dangling the sleeves of loose fitting garments over stoves. Fire safety education must stress the importance of the proper clothing during cooking.

- Seek legislative backing that would involve the enforcement community; they should participate in the widespread acceptance and installation of "smart stoves."

- Work with the insurance industry to create financial incentives for homeowners who install "smart stoves."

People with Disabilities

Educate the fire service and building design community on fire safety considerations for people with disabilities.

- Involve national disability groups in the education of the fire service, building industry, and design professionals as to the special needs of people with disabilities.

- Educate the fire service on the complications and challenges associated with the evacuation of people with disabilities from a private home or group home.

Organize disability and fire service representatives into a national coalition with two goals: to raise fire safety awareness among the disabled community, and to raise the awareness of the fire service to the needs of people with disabilities.

- Collect and review current educational materials for people with disabilities.

- Identify the gaps in the educational materials collected.

- Develop educational materials and training programs that target both the fire service and people with disabilities, and fill the gaps identified above. These programs may include video instruction (for those who learn by doing), distribution of facts and tips to disability sites on the Internet, and public service announcements.

- Raise awareness of the limitations of those with disabilities among building designers and managers.

- Use people with disabilities to develop training materials for apartment building managers, thus guiding their actions in the event of a fire.

- Create legal and financial incentives for designers to incorporate fire safety measures for people with disabilities into the design and construction of a building.

- Prepare and market disability awareness training, especially evacuation techniques, in National and State Fire Academy courses.

- Direct a portion of public education training to people with disabilities, focusing on topics such as fire sprinkler systems, specialized smoke alarms, utilizing areas of refuge, and encouraging people with disabilities to alert the fire department of their special needs prior to an emergency.

Improve emergency egress from buildings that house people with disabilities.

- Promote the redesign and engineering of current egress provisions.

- Expand accessibility standards to include appropriate evacuation procedures. This expansion would include designating areas of refuge to defend in place, developing appropriate detection alarms, and developing elevators that are safe for use during fires.

Form a coalition to expedite the implementation of the fire safe elevator.[11]

---

[11] A fire safe elevator is an elevator that can be used for safe egress during a fire emergency without firefighter control. The technology is available, it simply needs to be applied.

- Include representatives from national disability organizations, the American Society of Mechanical Engineers, elevator manufacturers and contractors, model codes and standard groups, and state, local, and federal regulatory agencies in the coalition.

- Devise an action plan with realistic and measurable goals for the design and implementation of the fire safe elevator.

- Collect and review current research, proposed studies, and those works already in progress.

- Convene the coalition to propose new codes for elevator standards.

- Collect literature on current elevator egress requirements.

- Seek funding for the coalition.

- Promote the coalition's involvement in the code process.

- Create a prototype elevator for use in a fire and test it for reliability, functionality, and cost feasibility.

Pay more attention to fire safety issues of people with disabilities during the code development and enforcement process.

- Disseminate information to disability groups and legislators about existing building codes as they pertain to people with disabilities.

- Improve data collection and research on the fire protection issue of people with disabilities.

- Promote necessary adaptations that will improve how building fire codes address the safety of people with disabilities.

- Promote code enforcement at the state and local levels.

Universal Messages

Inevitably, the three target groups at the conference overlapped in some of their ideas, particularly when discussing the need for home fire sprinkler systems. For example, the group targeting children stressed the need for fire sprinkler systems to combat the high rate of residential fires caused by children playing with fire starting materials. Fire sprinkler systems and fire resistant materials are particularly important safeguards to young children, especially when unsupervised; these fire safety features help protect kids who can't make decisions for themselves.

The group focusing on older adults stressed the need for fire sprinkler systems for the elderly because of the high number of fires caused by smoking materials and those caused by heating elements (space heaters and electric blankets, for example) among older adults. Fire sprinkler systems are also particularly beneficial for people with disabilities; the time it takes them to evacuate a fire may allow the fire to grow in size and intensity, resulting in death or injury. A fire sprinkler system could prevent a fire from reaching these dangerous stages. A compilation of ideas pertaining to more than one group is listed below.

Promote the installation of home fire sprinklers, make the costs more affordable, and educate the public on the benefits of fire sprinkler systems.

- Educate new home buyers and current home owners on the facts about home fire sprinkler systems. Fire sprinklers save lives, and it is in the home where 80 percent of all fire deaths occur. The goal is to increase the prevalence of home fire sprinkler systems to equal that of commercial fire sprinkler systems.

- Ask home builders to encourage new home buyers to install fire sprinkler systems in their homes, or at least provide information on the costs and benefits.

- Dispel the beliefs that if one fire sprinkler head activates, the whole system will activate, thus flooding the entire home. Each head is individually activated by the presence of heat.

- Work with industry to find ways to make home fire sprinkler systems more affordable.

- Emphasize the effectiveness of fire sprinkler systems. While fire extinguishers can often be used to mitigate a small cooking or trash can fire, a complete home fire sprinkler system will prevent a significant loss should the fire expand beyond the firefighting capabilities provided by a fire extinguisher. The risk for injury or death is greatly reduced when a residential fire sprinkler system is in place.

- Combine the expertise of groups such as National Fire Sprinkler Association (NFSA), American Fire Sprinkler Association (AFSA), the National Institute of Standards and Technology (NIST), and the fire service industry to collaborate on identifying current problems in home fire sprinkler systems, as well as to propose solutions for future systems.

- Create insurance incentives for homes and businesses that do invest in fire sprinkler systems.

- Continue research into the effectiveness of home fire sprinkler systems.

- Seek funding for independent agencies to conduct extensive research on home fire sprinkler systems.

- Publish data/findings of research; support the case to install a fire sprinkler system.

Form a coalition that will make fire safety a primary concern by raising our safety expectations for the environments to which our children, older adults, and people with disabilities are exposed.

Have the coalition assist in the following areas (which are also useful to do independent of any coalition):

- Educate home-buyers and renters to ask, as routinely as they would neighborhood crime, whether the house is fire-safe, especially with respect to members of the family who are children, older adults, or people with disabilities.

- Educate people to investigate schools, daycare centers, retirement homes, and centers specializing in the needs of people with disabilities to assess the fire safety measures in place to protect themselves and their loved ones, in addition to a well thought-out and rehearsed evacuation plan.

- Solicit grants and Federal funding to help reduce the price of fire detection, notification, and suppression devices/systems to allow the retrofitting of older buildings to become a reality.

- Advocate for effective legislation that addresses fire protection measures for people with disabilities.

Forming a coalition to address fire safety codes and standards must be acted upon swiftly. Representatives from the fire protection arena, NFPA, children's groups, older adult advocacy groups, and disability groups must unite to formulate a work plan. After the coalition's initial findings, lobbying at the local, state, and federal level must take place so the fire service can expect enforcement and compliance to the new codes and standards. The coalition should also seek out grant moneys to help offset the costs. Fire safety should be considered a must, not a crippling financial drain.

Mandate more built-in fire safety in new construction; it is less expensive to install fixtures to outfit a structure during its initial construction phase, than it is to later retrofit.

- Require more of the materials used to build new homes to be fire resistant.

- Mandate the installation of appropriate mitigation and early warning devices and components, such as commercial or home fire sprinkler systems, extinguishers, and smoke alarms appropriate to the population being served (for example, specialized smoke alarms are required for the deaf and hard of hearing).

ACTION TAKEN IN RESPONSE TO SOLUTIONS 2000 REPORT

The findings of Solutions 2000 need to be put into action to reduce fire-related injuries and deaths among those populations at greatest risk. We encourage you to implement the recommendations in this report and to share your efforts with us. Documentation will be made of all initiatives reported so that others might share and build on your successes.

What actions have been taken by your organization in response to the recommendations of Solutions 2000?

Please summarize each initiative by including the following:

- target group,

- recommendation implemented,

- partners or coalition members,

- procedure,

- results to date,

- method of evaluation planned or performed,

- plan for continuation,

- lead organization,

- contact person (name, title, mailing address, phone and fax numbers, etc.)

Include a copy of materials produced when possible. Return information prior to January 2001 to:

North American Coalition for Fire and Life Safety Education
c/o Peg Carson
Carson Associates, Inc.
35 Horner Street, Suite 120
Warrenton, VA 20186

APPENDIX A: "ENLIGHTENMENT AND EDUCATION"

– Philip Schaenman, President, TriData, Arlington, VA

Philip Schaenman, a former associate administrator of the U.S. Fire Administration, is a national spokesperson on fire prevention, innovative management practices, and new technology for the fire service. Mr. Schaenman has written and published various reports relating to public fire education, such as "Reaching the Hard to Reach" and "Proving Public Fire Education Works." In addition, Mr. Schaenman has published extensive reports examining international concepts in fire prevention. Mr. Schaenman was asked to address the benefits of and challenges in public fire education.

Public fire education is arguably the most productive aspect of fire protection. In terms of casualties and dollar loss saved, public education has proven its effectiveness. Studies in the United States and abroad have shown that emphasis placed on good public education has led to a decrease in fire deaths.

The central goal of public fire education is to change people's behavior:

- Teach how to prevent the most common causes of fire.

- Teach how to compartmentalize (e.g., by closing doors), especially for people with disabilities who cannot escape.

- Teach how to extinguish a fire and when to flee rather than fight.

- Teach when, how, where to escape; refuge sites.

- Teach how to report fires; early warning is especially vital for the elderly and people with disabilities.

- Teach the decision-making steps associated with dealing with a fire.

Public education is also useful in influencing fire safe engineering and increasing the public's use of various products.

- Inform the public about available fire protection technology they can use, such as alarms and home sprinkler systems.

- Dispel false ideas of fire sprinkler systems that prevent more widespread demand.

- Promote fire safe construction

- Inform the public of unsafe appliances, using CPSC information

By shedding light on unsafe practices and the rationale behind codes, public fire education assists fire code enforcement officials and increases the acceptability of fire safety standards. For example:

- Dispel marketing myths, such as reduced desirability of products with fire safety information because the seller thinks it scares the consumer and hence is bad for business.

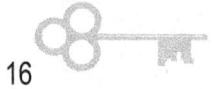

- *Building Codes.* Expose reductions in safety standards caused by weakening building code to entice new investors and new residents by making construction cheaper.

- *Code Enforcement/Arson Prevention Programs.* Teach the public why we have codes and the importance of adherence and compliance to them.

Why is the Fire Death Rate So Much Greater in the U.S. Than in Other Nations?

- *Fire Service Involvement.* Fire prevention in the United States is considerably different from that in Europe. More built-in protection, stronger code enforcement, and more public education may be found in many nations. Another major difference is the degree of participation of the fire service in prevention. In the United States the first priority of the fire service is suppression. Very few resources and personnel are allocated to public education efforts. In contrast is – or soon will be – the United Kingdom, where most members of the fire service are now expected to be involved in the delivery of public fire education. The fire service tailors prevention efforts to meet neighborhood needs, and extensive training on prevention is given to recruits and to junior officers in many nations. Higher levels of technical training are expected of fire officers.

- *Public Apathy.* Apathy towards fire prevention is prevalent in American society. The public tends to be unaware of the magnitude of the losses that are associated with fires, and of the huge total cost of fire (over $100 billion per year). The United States has not capitalized on the opportunity of using our fire experiences to teach messages of prevention. This is exemplified by the lessons learned from the Great Chicago Fire of 1871. The loss of 300 lives, 100,000 people left homeless, 18,000 buildings reduced to ashes, and $200 million in property damage has been reduced to the image of Mrs. O'Leary's cow. In the face of such pop culture reductionism, important lessons have been lost. We do not explain fire's role in our cultural history in schools, unlike the lessons taught in nations ranging from Japan to England.

  The gravity of other social ills detracts from fire problem awareness. This country is more entranced with crime and gives it disproportionately greater attention than its relative risk. In other instances, socially acceptable practices heighten the risk for fires. Alcohol has been a persistent threat to public health and deaths due to fires are highly associative with intoxication: about half of adult fire fatalities are legally drunk.

Keys to Successful Public Education

In general, the following are some key features of successful public education programs.

- *Market Research.* Once having identified a target audience, market research in some form is needed to tailor the programs to their intended audience and make sure they will understand and have impact. Public education programs need to be tested and refined before being implemented on a widespread basis. This is vital to avoid various cultural misinterpretations of the message.

- *Broad Outreach.* Once developed, a program must reach a significant portion, preferably the majority, of the intended population to have significant, bottom-line impact. This means more than implementing a few pilot projects.

- *Increased Use of the Fire Service and Partnerships.* Involve more of the fire service and partnerships to increase the outreach of public education, especially public education of high-risk groups. Leverage the fire service's limited resources with alliances with the local and national business community, and with special interest groups representing the targeted populations – which is the concept behind this conference.

Young Children

- Kids are curious. They are not aware of the power and effects of a fire and therefore must be taught. Teach parents what to teach kids, as well as the kids themselves.

- Use the caregivers for young children (e.g. hospitals for educating new mothers; people who run child care centers)

- Target low income families, single moms

- Videos are effective for reaching kids

Older Adults

- Piggyback on existing programs that have demonstrated their effectiveness

- Target the caregivers, depending on the type of residence. Train in-home nurses, deliveries of meals-on-wheels, etc.

- Come up with better ways to alert the fire department (medic alert tags, mobile phones).

- Entice the elderly to social events that discuss fire safety in addition.

- Grab their attention. Use what is important to older adults in your messages, e.g., their pets.

- Place the messages where older adults will see and hear them the most. You must use market research to determine these sites.

People with Disabilities

- Encompass a variety of disabilities, such as blindness, deafness, and mobility impairments. Each has its own problems in dealing with fires.

- Deal with detection, mitigation, and escape planning for each.

- Address the particular problem of each developmental and mental disability groups (e.g. the hearing-impaired may not hear an alarm, but can escape when they do. The mobility-impaired hear the alarm, but have difficulty escaping. People with even moderate mental impairments can learn basic skills through repetition and physical demonstrations).

APPENDIX B: "ENGINEERING THE ENVIRONMENT"

– Wayne "Chip" Carson, P.E., President, Carson Associates, Inc., Warrenton, VA

Wayne Carson is the President of Carson Associates, Inc., a consulting firm specializing in fire protection engineering and loss control. Mr. Carson has more than 25 years of experience in fire science technology and is an expert in the fields of fire code interpretation and analysis and fire protection system design. Mr. Carson was asked to address the physical characteristics of fire and the engineering approaches to circumvent the deleterious effects of fire.

What exactly is fire? Fire is the result of a series of chemical and physical reactions producing a variety of dangerous by-products, including

- Heat

- Smoke

- Other Toxic Products of Combustion

Heat from a fire quickly raises the temperature of all items in the room or space. In a very short time, a small fire can rapidly grow and cause what is know as a "flashover," in which all flammable items in a room reach their ignition temperature and burst into flames. This may occur even if an item never comes into contact with a spark or flame. Smoke, a second harmful by-product, serves to disorient the victim of a fire. It acts as an irritant to the eyes, causing blurring and tearing, as well as an irritant to the respiratory system. Smoke fosters fear, uncertainty, and panic. Unable to breath and unable to see, an individual's ability to make rational decisions is seriously diminished. Equally sinister are the toxic gases that are produced in a fire. For example, carbon monoxide is an odorless, colorless, and tasteless gas that produces most of the deaths in fires.

The goal of fire protection engineering is to control or reduce the development of a fire in an attempt to provide occupants more time to escape. The development of a fire follows a series of phases, the first of which is the growth period in which heat released from the fire gradually raises the temperature of all items in the room. In a matter of minutes, all combustible materials will be heated to the temperature at which they spontaneously ignite. The room is now totally involved in fire. When the fire has consumed all the combustible materials, it enters the decay period, and the temperature gradually decreases.

As a fire develops, a sequence of events occurs that contributes to whether a person will be able to detect, mitigate, and escape the fire danger. The first of these steps involves detection, followed by alarm notification, response, and extinguishment. Each of these actions take time. The goal of fire safety is to reduce this time and make sure the occupant can evacuate to a safe place, normally outside, before the fire produces untenable conditions.

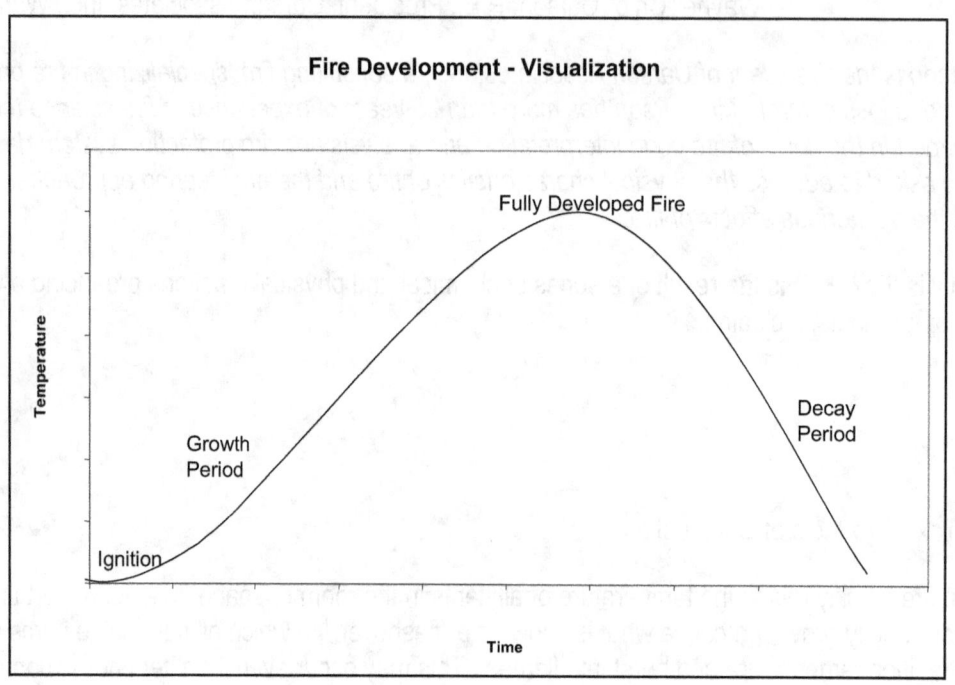

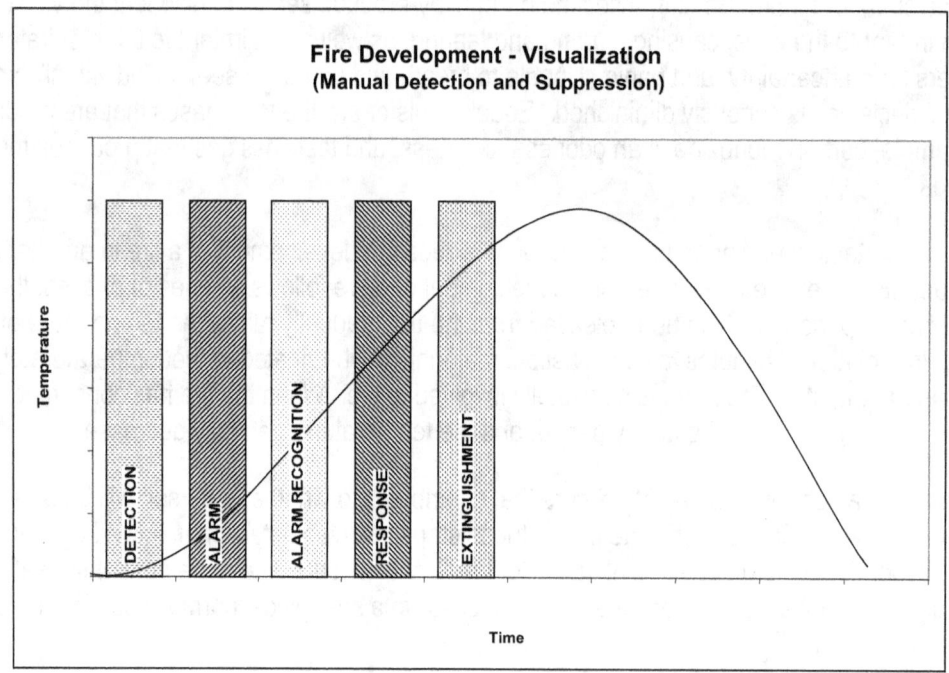

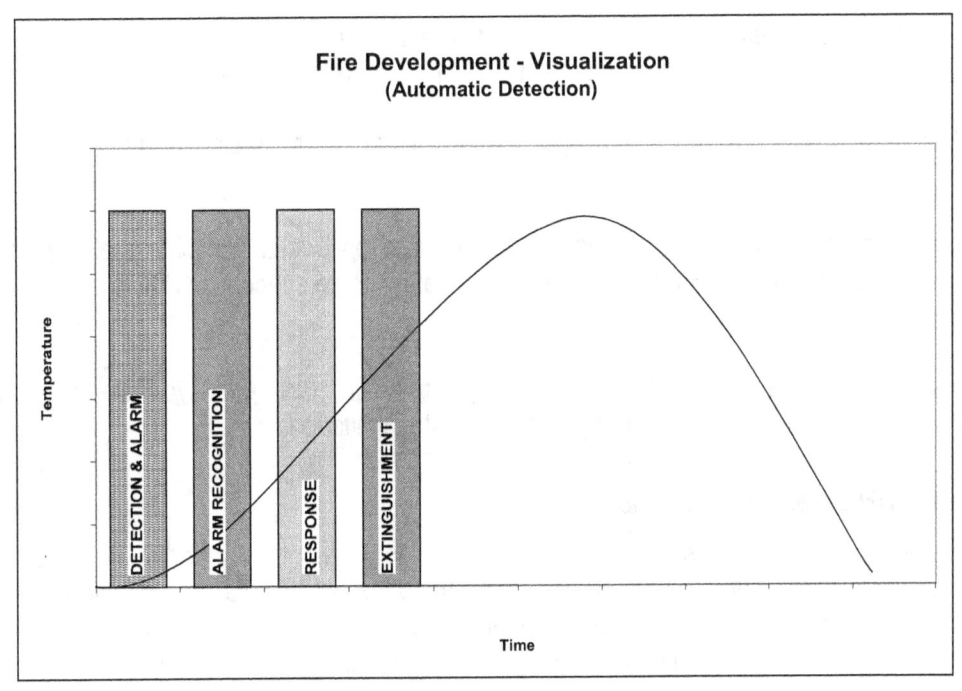

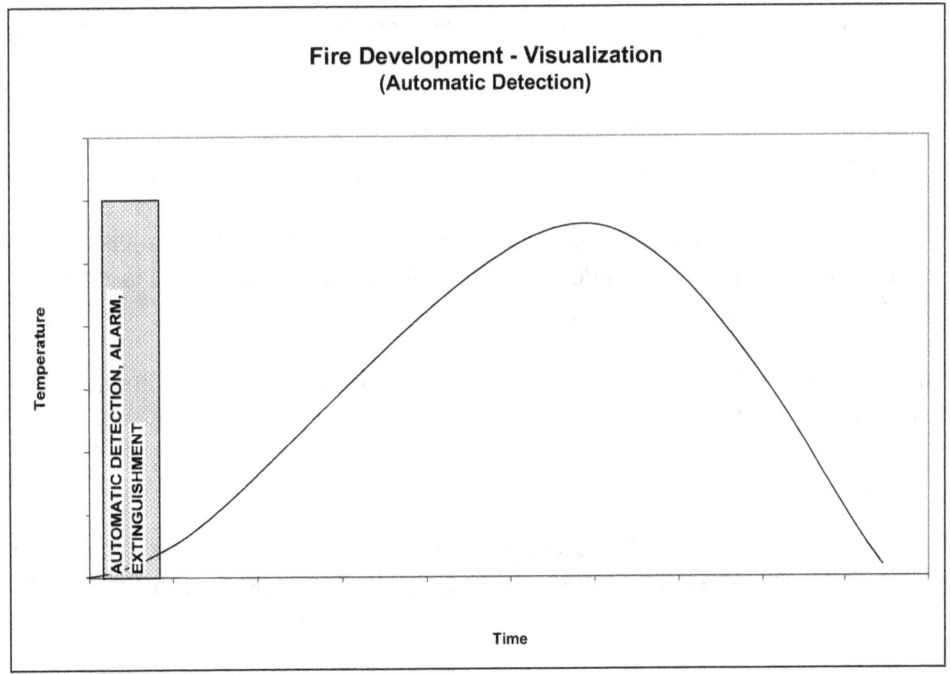

Reliance on human detection and manual suppression alone can significantly increase escape time in the event of a fire. Under the best of conditions, one has only few precious minutes to safely evacuate a burning structure. Flashover can occur in as little as 2-4 minutes, but smoke and gases can kill or incapacitate in even less time. Manual detection, alarm, and fire suppression can cause significant time delays, thus placing the occupants at greater risk for death or injury. Delays from manual actions affect every phase of a fire event:

- *Detection.* Delay in time if fire is not immediately recognized

- *Alarm.* Possible delay before sounding alarms or failure to sound altogether. People often waste time investigating the alarm or trying to extinguish a fire before sounding the alarm.

- *Alarm Recognition.* Delayed because people often assume it is a false alarm or not an immediate threat.

- *Response.* Fire department response may be delayed due to weather, traffic, time of day, or availability of closest unit. Occupants may not immediately begin evacuation due to a variety of reasons.

- *Extinguishment.* Fire department attempts to extinguish the fire may be delayed due to time necessary to lay hoses, locate and connect to fire hydrants, and staff limitations.

Automatic Detection and Suppression

An automatic detection and suppression system dramatically increases the chance of escaping a fire in a timely manner. Installation of an automatic detection system affords rapid alarm recognition and notification, and subsequent evacuation. Early extinguishment by an automated system, such as fire sprinklers, can control the fire growth and possibly eliminate the need for evacuation.

- *Detection and Alarm.* Automatic alarm condenses two steps (detection and alarm) into one and saves time.

- *Alarm Recognition.* Though still not problem-free, recognition is faster with automatic suppression/detection systems as the alarm is activated sooner.

- *Response.* Initiates a more rapid fire department response as the fire department is automatically notified upon fire detection. Also, provides occupants with an early warning, signaling the need to evacuate before the fire and smoke prevent escape.

- *Extinguishment.* Early suppression prevents flashover and limits fire growth. Therefore, the amount of smoke and heat are reduced.

Protection Methods

There are a number of everyday things that consumers can do to minimize their risk of becoming a casualty of fire. The following are examples of simple engineering and consumer product safety approaches to fire protection:

Fuel Control

- Housekeeping – Get rid of the clutter and combustible materials; a smaller fuel load will result in a less devastating fire should ignition occur.

- Clothing – Wear tight-fitting clothing while cooking. For example, loose fitting sleeves may easily contact open flame or heating elements. Fire resistant sleep-wear for children reduces the likelihood of clothing fires.

- Interior finish – Reduce flammable material on walls and ceilings and use flame resistant materials to slow down fire development.

- Furnishings – Primary contributing factor to the growth of a fire; select for fire resistance and slow flame spread.

- Hazardous materials – Gasoline, chemicals, etc.; store in proper containers, store outside or in a garage, if possible, and use according to instructions.

- Building construction – Compartmentation limits the spread of fire and smoke. Simply closing doors can slow fire and smoke spread.

Ignition Control – Build in Safety and Maintain

- Heating equipment

- Electrical equipment and appliances

- Cooking equipment

- Matches and lighters

- Use public education for proper use of ignition materials

- Maintain equipment and systems

Building Arrangement-Construct According to Codes and Maintain

- Egress

- Exits (outside at grade)

- Areas of refuge

- Compartmentation

- Corridors

- Living Units

- Smoke Barriers

Fire Detection and Suppression-Install and Maintain

- Fire detection

  - Smoke detectors for early warning

- Alarm notification

  - Audible

  - Visual

- Automatic fire department notification

Fire Sprinklers

- Complete throughout building

- Quick response fire sprinklers for earlier activation

Choose the Best Options for Relevant Conditions

When developing fire protection for a building, whether built-in construction or products for consumer use, one must consider the conditions surrounding the installation and use of that product. Some things to consider are:

- New vs. existing structures – must create products that can be retrofitted

- High rise vs. low rise structures

- Historic preservation

- Cognitive capability of building occupants

- Evacuation capability of building occupants

- Staff support available to occupants, such as in congregate living homes

APPENDIX C: "ENACTMENT AND ENFORCEMENT"

– Mary L. Corso, Washington State Fire Marshal, Olympia, WA

Mary L. Corso is the State Fire Marshal for the State of Washington and is one of the first women to hold such a position in the United States. She has more than 22 years experience in the fire service and is an expert in the field of fire protection enforcement.

Who is responsible for fire safety enforcement? Effective enforcement requires a collaborative effort between enforcement agents, those who are responsible for providing fire protection, the community, and government regulators. The community, largely the fire service, is responsible for addressing the fire safety needs of the populations residing within. Providers are responsible for knowing codes and regulations and understanding their benefits. Government regulators are responsible for setting standards meant to ensure a safe environment. Lastly, it is the responsibility of the enforcement community to see that these standards are being followed. However, it is important that each group understands the unique issues of this vulnerable population.

Given the complexity of participants, a variety of attitudes come into play when attempting to enforce an issue. Inevitably, the differences in opinion lead to barriers that can hamper the enforcement processes. The attitudes of many fire code enforcers tend to be simple, and as a result, often unrealistic. Problematic thinking on the part of fire safety enforcers is that fire safety is all that matters and cost should not be a factor. Cost is arguably the primary reason behind any hesitation on behalf of those responsible for fire safety provisions. One size does not fit all. Strict adherence to a fire code or regulation may cause unnecessary burden on the part of providers and at the same time do little to raise the fire safety level. Behavioral patterns may be as much to blame for fires as is faulty construction and non-code compliance. Smoking fires can be eliminated if we simply prohibit or monitor all smoking, but this idea is as ludicrous as it sounds in a nation that deeply cherishes individual freedom. Some enforcers do not realize that their role is ongoing and needs continuous evaluation. Fire sprinkler systems may be the most effective means for preventing fire deaths, but it is not enough to simply sprinkler it and forget it. It is foolish to fire sprinkler a building that will be torn down in the next year. In such cases, enforcers must remember the spirit of the law and apply common sense to developing alternative measures. What's more, just adding more staff is not enough to achieve widespread and thorough enforcement. The field of fire safety enforcement is complex and requires critical thinking and constant re-evaluation.

The attitude of the fire safety provider can be equally unyielding and may often act as a barrier to change. Many providers erroneously think it costs too much money for fire safety and that upgrading may put them out of business. Several not-for-profit organizations feel that they are exempt from expensive upgrades. In other cases, fires may happen too infrequently to warrant a change. Many businesses have been in operation for years and have never experienced a fire.

Group Assisted Living Homes

In one year alone, the state of Washington experienced 11 fire deaths in residential group living facilities. Much like the rest of the country, fire deaths in this state are declining, however specific incidents such as group living home fires are on the rise. Fires of this nature present a significant challenge to all parties involved in improving fire safety through enforcement. It is the responsibility of group home facilities to provide a safe living environment for its residents, however a myriad of circumstances unique to group assisted living homes must be examined before implementing a fire protection system.

Provider Issues

- Aging-in-place; how do retirement homes and nursing facilities cope with progressively deteriorating individuals with regard to fire safety; do they have adequate staff to evacuate all occupants in a reasonable time frame?

- Choice (e.g., smoking vs. non-smoking facilities)

- Residents rights

- Evacuation requirements – resident capabilities change day by day

- Staffing ratios – night/day variance

- Codes that solve one problem, yet create another

- Funding overall

Enforcement Issues

- Evacuation criteria

- Placement of occupants relative to exits and evacuation tools, such as ramps and mechanical chairs for going up or down stairs, etc.

- Staffing

- Protection features – existing facilities

- Grandfathering

- Code adoption process

- Codes/standards/regulations may be conflicting

- Lack of enforcement resources – especially problematic for volunteer fire departments

Evacuation Issues

- Defend-in-place

- Sheltering

- Aging-in-place; some people are completely incapacitated

- Performance based

Code Issues

- Standardization and gaps in enforcement

- Adult family homes are largely unregulated for fire safety

- Movement from higher level of protection into family environments; increases quality of life, but creates a high risk environment.

- Codes often must be written in blood. Death begets change as fatal fires provide a rare window of opportunity to improve enforcement.

- Codes are often myopic – "One size fits all" syndrome makes it difficult to remember the spirit and intent of the code. Negates common sense.

Fire Protection Issues

- Staffing ratios/supervision

- Resident placement

- Ignition control

- Fuel control

- Building arrangement

- Staff training

- Disclosure statements

- Enforcement

Benefits of Protection

Innovative measures aimed at increasing enforcement emphasize the benefits of protection to providers. Among these are financial incentives, legal incentives, and better public relations. Expanding fire safety measures in group assisted living homes can increase the occupant load and use of the facility. Installing more efficient fire safety measures may save money in the long run by doing away with more costly measures already in place. For example, automatic fire sprinklers and designated areas of fire refuge will reduce the staff size that would be necessary to evacuate all residents in the event of a fire. In addition, facilities with improved fire safety mechanisms may receive substantial discounts from insurance companies. The benefits of complying with the law are self-explanatory. Facility owners can look forward to reduction in fines and penalties caused by poor code compliance. Better fire safety makes for better business. Not only will you have a building that will last for generations, but public opinion will be favorable and promote use of the facility.

Future Issues

As we move into the next century, the enforcement community must address demographic shifts, changes in residential home occupants, and scant resources. The population is aging and the number of US citizens over the age of 65 is expected to rise exponentially when the Baby Boom generation reaches the retirement age. We can expect to see continued movement of older adults into group home environments. Coupled with aging structures, the increase in group home residents will place a significant portion of the population at serious risk. Using a systems approach, enforcers, providers, regulators, and the community can start today to build a safer tomorrow.

APPENDIX D: SOLUTIONS 2000 PARTICIPANTS

1. Aguilar, Carmen
   Deaf Services Advocate
   National Center for Latinos With Disabilities
   1921 S. Blue Island Ave.
   Chicago, IL 60608
   800/532-3393 (V/T) or 312/666-3393 (V)
   312/666-1788 (TTY) or (F) 312/6661787
   ncld@ncld.com

2. Albarelli, Lois
   Administration on Aging
   Aging Services Program Specialist
   Room 4748 Cohen building
   330 Independence Ave. SW
   Washington, DC 20201
   202/619-2621 (F) 202/260-1012
   Lois.Albarelli@aoa.gov

3. Amiri, Shahriar, CBO
   Division Chief, Building Construction
   Montgomery County Department of Permitting
   Services
   255 Rockville Pike, Second Floor
   Rockville, MD 20850-4166
   301/217-6224 (F) 301/217-6381

4. Appy, Meri-K
   Vice President of Public Education
   National Fire Protection Association
   1 Batterymarch Park
   Quincy MA 02269
   617/984-7288 (F) 617/770-0200
   mappy@nfpa.org

5. Bassett, Gerry
   Program Chair, Education
   United States Fire Administration
   16825 South Seton Ave.
   Emmitsburg MD 21701
   301/447-1094 (F) 301/447-1178
   gerry.bassett@fema.gov

6. Beale, LaTanya
   Deputy Director, NCIPC – CDC
   Division of Unintentional Injury Prevention
   4770 Buford Hwy, NE, Mailstop K63
   Atlanta, GA 30341
   770/488-4652 (F) 770/488-1317

7. Beasley, Kim
   Director of Architecture
   PVA
   801 Eighteenth St., NW
   Washington, DC 20006
   202/416-7644 (F) 202/416-7647

8. Berger, Helena
   Chief Operating Officer
   American Association of People With
   Disabilities
   1819 H St., NW, Suite 330
   Washington DC 20006
   202/457-0046 (F) 202/457-0473
   Hberger952@aol.com

9. Black, Brian
   Director, Building Codes & Standards
   Eastern PVA
   111 West Huron St.
   Buffalo, NY 14202
   716/856-6582 (F) 716/855-3395
   bdblack55@aol.com

10. Brown, Robert J., CBO
    Program Manager
    International Code Council, Inc.
    5203 Leesburg Pike, Suite 708
    Falls Church VA 22041
    703/931-4533 (F) 703/379-1546
    brown@intlcode.org

11. Bryant, William
    Code Enforcement Administrator
    PACE
    Inspections and Environmental Programs
    2664 Riva Road
    PO Box 6675
    Annapolis, MD 21401
    410/222-7737 (F) 410/222-7970

12. Burns, James A.
    State Fire Administrator
    NYS Department of State
    OFPC/12th Floor
    41 State Street
    Albany, NY 122331-0001
    518/474-6746
    jburns@dos.state.ny.us

13. Carson, Chip
    Carson Associates Inc.
    35 Horner St., Suite 120
    Warrenton, VA 20186
    540/347-7488 (F) 540/349-9147
    carsonfpe@aol.com

14. Carson, Peg
    Carson Associates Inc.
    35 Horner St, Suite 120
    Warrenton, VA 20186
    540/347-7488 (F) 540/349-9147
    carsonpeg@aol.com

15. Catanzaro, Peter
    Fire & EMS Specialist
    Ogilvy Public Relations Worldwide
    1901 L St, NW, Suite 300
    Washington, DC 20036
    202/452-9508 (F) 202/331-3003
    peter.catanzaro@dc.ogilvypr.com

16. Charpentier, Teri
    American Red Cross of Central Mass.
    Juvenile Firesetters Program
    61 Harvard St.
    Worcester, MA 01613
    508/756-5711x316 (F) 508/793-8621
    charpent@usa.redcross.org

17. Cochran, John
    Fire Management Specialist
    United States Fire Administration
    16825 South Seton Ave.
    Emmitsburg MD 21727
    301/447-1421, (F) 301/447-1102
    john.cochran@fema.gov

18. Corso, Mary
    Washington State Fire Marshal
    Fire Protection Bureau
    11th and Columbia/ PO Box 42600
    Olympia, WA 98502
    360/753-0404 (F) 360/753-0398
    mcorso@wsp.wa.gov

19. Cote, Arthur
    Sr. Vice President Operations
    National Fire Protection Association
    1 Batterymarch Park
    Quincy MA 02269
    acote@nfpa.org

20. Dalton, Jim
    Planning Team/Coalition
    Director of Public Fire Protection
    National Fire Sprinkler Association, Inc.
    P.O. Box 810
    Warrenton, VA 20188
    540/937-3466 (F) 540/937-3466
    Dalton@nfsa.org

21. Dedman, Jackie
Associate Director
Disability Services Quality Improvement Center
(DSQIC)
University of Arkansas For Medical Science
501 Wood Lane, Suite 210
Little Rock AR 72201
800/831-4827x9911 or 501/682-9900
(F) 682-p9901
dedmanjacqualin@exchange.uams.edu

22. Endthoff, Gene
National Fire Sprinkler Association, Inc.
429 South Locust
Sycamore, Illinois 60178
815/895-5521 (F) 815/899-5521
Endthoff@nfsa.org

23. Erdheim, Rick
Senior Manager, Government Affairs
National Electrical Manufacturers Assoc.
1300 N. 17th St., Suite 1847
Rosslyn, VA 22209
703/841-3249 (F) 703/841-3349
ric_erdheim@nema.org

24. Fellner, Hank
Fire and Burn Team Leader
The National SAFE KIDS Campaign
1301 Pennsylvania Ave. NW
Suite 1000
Washington DC 20004
202/662-0621 (F) 202/393-2072
Hfellner@safekids.org

25. Gamache, Sharon
Executive Director
NFPA Center For High Risk Outreach
1 Batterymarch Park
Quincy MA 02269
617/984-7286 (F) 617/770-0200
sgamache@nfpa.org

26. Gratton, Jan
International Fire Marshals Association
469 S Albertson Ave
Covina CA 91723
626/966-8070 (F) 626/966-8721
jangratton@msn.com

27. Gundersen, G. Mark
Associate
Ogilvy Public Relations Worldwide
1901 L Street, NW
Suite 300
Washington DC 20036
202/452-9449 (F) 202/331-3003

28. Hall, John
Assistant Vice President
Fire Analysis and Research
1 Batterymarch Park
Quincy MA 02269
617/984-7460 (F) 617/984-7478
jhall@nfpa.org

29. Harp, N'ann
President
Smart Consumer Services
2111 Jeff Davis Highway, Suite 722 North
Crystal City, VA 22202
703/416-0257 (F) 703/416-0258
nannharp@aol.com; sconsumer@aol.com

30. Harper, Leon
Manager of Housing Programs
AARP
601 E St., NW
Washington, D.C. 20049
202/434-6049 (F) 202/434-6466
lharper@aarp.org

31. Harvey, Pauline
Public Health Advisor
NCIPC – CDC Division of Unintentional Injury
Prevention
4770 Buford Hwy, NE, Mailstop K63
Atlanta, GA 30341
770/488-4652 (F) 770/488-1317

32. Herman, Andrea
Vice President, Communications
Sleep Products Safety Council
501 Wythe St.
Alexandria, VA 22314
703/683-8371 (F) 703/683-4503
aherman@sleepproducts.org

33. Hoebel, James F.
Chief Engineer for Fire Safety
US Consumer Production Safety Comm
4330 East-West Highway
Bethesda, MD 20814
301/504-0494x1380 (F) 301/504-0533
jhoebel@cpsc.gov

After June 3, 1999
13506 Star Flower Court
Chantilly, VA 20151
703/818-2639
jfhoebel@erols.com

34. Hogan, Erin
Research Associate
TriData Corporation
1000 Wilson Boulevard, 30th Flr
Arlington, VA 22209-2211
703/351-8300 (F)703/351-8383

35. Isman, Ken
National Fire Sprinkler Association, Inc.
Robin Hill Corporate Park
Route 22 – PO Box 1000
Patterson, NY 12563
914/878-4200 (F) 914/878-4215
Isman@nfsa.org

36. Juillet, Edwina
Consultant, Fire & Life Safety for People With
Disabilities
Egypt Bend Estates
637 Riverside Drive
Luray, VA 22835-2910
804/243-6353 (F) 804/982-0821
edwina@shentel.net

37. Katcher, Dr. Murray
University of Wisconsin Medical School
        Dept. of Pediatrics, H6 /440
600 Highland Ave.
Madison, WI 53792
608/262-8416 (F) 608/263-0440
mkatcher@facstaff.wisc.edu

38. Keith, Gary, AVP
Director Regional Operations
National Fire Protection Association
1 Batterymarch Park
Quincy MA 02269
617/984-7260 (F) 617/984-7110
gkeith@nfpa.org

39. King, William
US Consumer Productions Safety Commission
4330 East-West Highway
Bethesda, MD 20814
301/504-0494x1296 (F) 301/504-0533
wking@cpsc.gov

40. Little, Leslie
Help-U
1409 B North Mount Vernon Ave.
Williamsburg, VA 23185
757/221-0542 (F) 757/221-8377
helpu@visi.net

41. Lopes, Rocky
American Red Cross
Community Disaster Education
8111 Gatehouse Rd, 2nd Flr
Falls Church VA 22042
703/206-8805 (F) 703/206-8848
lopesr@usa.redcross.org

42. Marx, Toney
National Vice President
PVA
849 70th Ave., SE
Salem, OR 97301
503/585-9324 (F) 503/585-3085

43. Mathis, Chris
    Director, Building Technologies
    Smart Consumer Services
    2111 Jeff Davis Highway, Suite 7227 North
    Crystal City, VA 22202
    703/416-0257 (F) 703/416-0258
    rcmathisl@aol.com

44. Mickalide, Angela
    Program Director
    The National SAFE KIDS Campaign
    1301 Pennsylvania Ave. NW
    Suite 1000
    Washington DC 20004
    202/662-0603 (F) 202/393-2072
    amichalide@safekids.org

45. Morris, Rich
    Co-Chair
    Ontario Fire Marshal's Public Fire Safety Counci
    50 East Pearce
    Richmond Hill Ontario L4B1B7

46. Muncy, Steve
    President, American Fire Sprinkler Association
    12959 Jupiter road, Suite 172
    Dallas, TX 75238
    214/349-5965 (F) 214/343-8898

47. Neal, Wayne
    EMSC National Resource Center
    111 Michigan Ave., NW
    Washington DC 20010
    301/650-8281 (F) 301/650-8045

48. Nickson, Ron
    Vice President Building Codes
    National Multi Housing Council
    1850 M Street, NW, Suite 540
    Washington, DC 20036-5803
    202/974-2327 (F) 202/775-0112
    rnickson@nmhc.org

49. Nolan, Steve
    State Farm Insurance Companies
    Corporate Headquarters
    One State Farm Plaza D-1
    Bloomington IL 61710-0001
    309/766-7635,(F) 309/766-9173
    steve.nolan.aqha@statefarm.com

50. Ottoson, John
    Fire Data Specialist
    United States Fire Administration
    16825 South Seton Ave.
    Emmittsburg, MD 21701
    301/447-1272, (F) 301/447-1102
    john.ottoson.@fema.gov

51. Pecht, Jim
    Accessibility Specialist
    U.S. Architectural & Transportation Barriers
    Compliance Board
    (Access Board)
    1331 F St., NW, Suite 1000
    Washington, D.C. 20004-1111
    202/272-5439 (F) 202/272-5447
    pecht@access-board.gov

52. Placzankis, Donna
    Program Manager
    Disaster Planning For The Elderly
    American Red Cross
    3747 Euclid Ave.
    Cleveland OH 44115-2501
    216/431-3010x2157 (F) 216/431-3360
    placzand@usa.redcross.org

53. Preede, Kenneth
    Seniors Policy Analyst
    American Seniors Housing Association
    National Multi Housing Council
    1850 M St., NW, suite 540
    Washington, DC 20036-5803
    202/974-2300

54. Sanders, Russ
Metro Chiefs Liaison
NFPA Central Regional Manager
3257 Beals Branch Road
Louisville KY 40206
502/894-0411 (F) 502/894-0519

55. Sanders, Walter A.
Couselor to the Chairman
U.S. Consumer Production Commission
4330 East-West Highway
Bethesda, MD 20814
301/504-0213x2234 (F) 301/504-0768
wsanders@cpsc.gov

56. Santiago, Ray (Captain)
NJ Representative
National Association of Hispanic Firefighters
Camden City Fire Department
P.O. Box 1534
Camden, NJ 08105
609/757-7527 (F) 609/757-7243

57. Schaenman, Phil
President, TriData Corporation
1000 Wilson Boulevard, 30th Flr
Arlington VA 22209
703/351-8300 (F) 703/351-8383
pschaenm@sysplan.com

58. Schiefer, Mark
KIDDE International
Vice President of Marketing
1394 South Third St.
Mebane, NC 29302
919/563-5911x342 (F) 919/5632711

59. Schofield, Mark
AVP, Manager – Loss Prevention Education
Factory Mutual Engineering Corporation
1151 Boston-Providence Turnpike
PO Box 9102
Norwood, MA 02062
781/255-4627 (F) 781/255-4184

60. Schoonover, Ken, P.E.
Vice President, Codes and Standards
Building Officials & Code Administrators
4051 West Flossmoor Rd.
Country Club Hills, IL 60478
708/799-2300 (F) 708/799-0320
kschoono@bocai.org

61. Siegfried, Tom
Executive Director
Sharel Stokes Fire Sprinkler Public Education
Foundation, Inc.
135 Spring Isle Trail
Altamonte Springs, FL 32714
407/788-8873 (F) 407/788-0277
escgtlsig@aol.com

62. Slye, Loren
Representative for British Columbia Canada
Life and Fire Safety Education Council
Captain – Community Relations
Richmond Fire Rescue
#1 Fire Hall
6960 Gilbert Rd
Richmond, BC V7C 3V4
604/303-2715 or 278-5131 (F) 604/278-0547

63. Smead, Ellen
Consumer Coalitions Coordinator
Consumer Federation of America
1424 16th St., NW, Suite 604
Washington, DC 20036
202/387-6121 (F) 202/265-7989
esmead@essential.org

64. Stewart, David
PAIR Program Advocate
Litton Building
1207 Quarrier St, 4th Flr
Charleston, WV 25301
304/346-0847 (F) 304/346-0867

65. Summey, Doris P.
US Administration on Aging
Cohen Building
330 Independence Avenue, SW
Washington, DC 20201
202/619-3775 (F) 202/260-1012
doris.summey@aoa.gov

66. Taft, Mandy
Research Assistant
The National SAFE KIDS Campaign
1301 Pennsylvania Ave., NW, Suite 1000
Washington DC 20004-1707
202/662-0630 (F) 202/393-2072
 mtaft@safekids.org

67. Waddell, Bob
EMSC NRC
111 Michigan Ave., NW
Washington, D.C. 20010-2970
301/650-8067 (F) 301/650-8045
rwaddell@emscnrc.com

68. Watts, Bettye
Arkansas Department of Public Health
Division of Child Adolescent Health
Director, Fire Related Burn Prevention
Slot 17, 4815 W. Markham
Little Rock, AR 72205-3867
501/661-2718 (F) 501/661-2992
bwatts@mail.doh.state.ar.us

69. Williams, James
Research Associate
TriData Corporation
1000 Wilson Blvd., FL 30
Arlington, VA 22209
703/351-8294 (F) 703/351/8383
jwilliams@sysplan.com

70. Williams, Soy
International Code Council, Inc.
Government Relations Director
5203 Leesburg Pike, Suite 708
Falls Church, VA 22041
703/931-9475x11 (F) 703/379-1546
williams@intlcode.org

71. Wilson, Kim
American Red Cross
707 North Main Street
Wichita KS 67203
316/268-0884 (F) 316/268-0887

Group Moderators

Young Children: Hank Fellner, The National SAFE KIDS Campaign
Older Adults: Leon Harper, AARP
People with Disabilities: Kim Beasley, PVA

Group Facilitators

Provided by AARP:

Young Children: Rob Cryer
Older Adults: Cindy Shearin
People with Disabilities: Cal Broughton

Solutions 2000: Advocating Shared Responsibilities for Improved Fire Protection